Can You Hear It, Can You See It

A Guide to Becoming A Better Sound Pro

Henry Tucker

For Information, Contact
Leading Through Living Community, LLC
228 Auburn Avenue
Atlanta, GA 30303
info@leadingthroughliving.com
ISBN-13: 978-1-949266-05-4 E-book
ISBN-13: 978-1-949266-04-7 Paperback

For the
"Sound Guys and Gals"
who want to be the
best techs in the industry.

CONTENTS

PREFACE

Since I was a child, I have always loved music. I loved the idea of sound and music, and how sound was translated into music. At the age of six, I began taking five-gallon buckets and turning them into drums. For some reason the tempo of the song (or the flow of the music) just seemed to come naturally, even at that very young age. Amazingly, I never knew where this love and passion for music came from.

Growing up, my mother had an old 8-track player with a turntable top. She would always tell us not to touch her radio, but when she wasn't home, I would turn it on and play music from gospel to country, R&B to jazz. I always searched for ways to make what I heard sound better, make it flow.

After being a percussionist for years, opening up for many different choirs, playing with a variety of musicians that included gospel, jazz, and rap; I felt like I had more to offer that was untapped, like it was trapped inside, fighting to break free.

Under the leadership of my godfather the late Dr. Dennis Hagins, Sr., I was connected to a sound engineer by the name of

Skip Winsek who took me under his wing. He cultivated the little knowledge I had with the basics of everything I needed to know. Because I had a passion and a love for sound, I didn't just stop there. I began buying books, attending workshops and online seminars, and going to concerts to meet with the sound engineers backstage. I volunteered my services, helping set up staging and lighting. One of the most valuable pieces of wisdom that I learned from Skip is that "you will never understand how to troubleshoot a sound system if you don't understand how one is put together." I never forgot that, and I dreamed of starting my own audio production company.

This eventually came to fruition, and I launched my first show. Because it was done outside, I named my company Open Sound Productions. From that time until present day, we have been hired to do sound for town halls, galas, and outdoor art festivals. We have also done concerts, banquets, jazz shows by the pool, and even a few on the lake.

Open Sound Productions not only does sound, we also do workshops, trainings, consultations, and installations. My motto for the company has always been "We make it sound better."

By the time you finish this manual, you will understand my intention to help you improve upon what you already know and get compensated for it, without recreating the wheel. With over 20 years of experience, I want to encourage, cultivate, and inspire better sound technicians and sound operators. I also strive to help Ministers of Music, Pastors, and individuals who aspire to become sound professionals (and how to identify good ones).

CHAPTER 1
WHAT IS SOUND?

Everyone's interpretation of the term "good sound" is different, but the meaning of quality should be the same: **Excellent**.

Sound is created by "vibrations that travel through the air or another medium and can be heard when it reaches a person's ear". Sound is produced by continuous and regular vibrations, as opposed to noise.

CHAPTER 2
WHAT AFFECTS SOUND?

As you go through each venue and consider the event, you should always keep in mind that your job is to always to provide the best sound experience - the best QUALITY of sound - that you can with the knowledge you possess. Everything must be taken into consideration, from the size of the room and number of people who will be in attendance, to the air conditioner and other "seen yet unseen" noise contributors.

Sound can be changed by the padding of the pews, thickness of the carpet, type and quality of the drywall and windows, and even the height of the ceilings. These elements can play a vital role in the quality of the sound that is being reproduced.

Sound in Action - Real World Example

One of the hardest challenges of being a sound technician (sound tech) is being responsible for ensuring great sound in challenging venues, particularly venues that were never built to be event or church spaces. For example, most strip malls and store fronts were built to house businesses such as grocery stores and clothing departments. They were not designed to hold weddings, concerts, or religious services. Buildings without carpet and non-padded seating are the hardest to produce quality sound because the sound travels everywhere - there is nothing to absorb it (other than human bodies). Honestly, unless the event or organization leader is willing to invest in upgrades to the flooring, seating, insulation, and sound absorption pads, there is not much you can do as the sound tech to improve how things sound. Just stay positive, be as helpful as possible, and hope the sound is not too loud.

As your knowledge of sound increases, you will face more challenges. Stay calm and take it in stride.

You will arrive at a venue, take one look, and your heart will sink as your mind shouts the words, "Man, were they even THINKING about sound when they built this?!" Take a deep breath.

I am telling you now: some buildings are going to be very hard to produce quality sound in, but it's not impossible. Whitney George said, "You won't be judged by the situation you've been given, but rather by the way you handled it." You will possibly see me refer to this statement again throughout this guide. The following information takes you through how to overcome hardship.

CHAPTER 3
THE RESPONSIBILITIES OF THE SOUND TECHNICIAN

As the Sound Tech or Sound Operator, you are in control of the fundamental communication vehicle that delivers a potentially life changing message to a lost soul. This is a serious responsibility, and should never be taken likely.

Sometimes the sound technician's leadership responsibilities can cause irritation, and clients will disagree with your decisions. They may complain that the music is too loud, and at other times, the complaint will be that they are unable to clearly hear the music (the background singers were drowned out, the monitors were up too loud, the subwoofers were "too bassy", etc.). Do not get defensive. Remain calm, breathe, and remember you have a job to do.

As the Sound Tech, you NEVER want to create issues with the people you are serving. That being said, it does not hurt to entertain criticism - make adjustments when and if possible. And if they can't be made, explain why, then carry on.

Sometimes the problem is the system, but remember: one of the greatest tools you have as a Sound Tech is knowing the system inside and out. That way, if the system goes crazy, you know why, and you know how to fix it.

This is your business and your profession. Do not rely on assistants or helpers to shoulder the responsibility for things - right or wrong. You are being compensated to provide an exceptional sound experience. YOU. Good, bad, or indifferent, it all begins and ends with you. One of the greatest attributes you can have as a technician is to be punctual. In this field being on time is highly respected and expected. It is not a good look for you or your company to be labeled as "always late" or "never on time".

Remember the old cliché: TIME IS MONEY. Keep in mind promoters and coordinators set schedules and make moves according to Time. **We are all hired (or asked/volunteered) to do a service.**

Plan to be on time or early. It's a great component of excellent service.

CHAPTER 4
CHARACTERISTICS & QUALIFICATIONS
OF A SOUND TECH OR SOUND OPERATOR

The qualifications and characteristics required of a Sound Tech depend on the organization that needs him/her. Much of my experience is related to religious organizations, and the range varies from "beginners welcome" to "experts only". Truth be told, that range applies to many companies, too. So below, I have put together a few things that come up consistently.

Candidates must be coherent and responsive. They must also be willing. This is normally the standard in most small organizations. This is because in small companies and churches, it is not always easy to find qualified sound operators. They do not have the budget to hire someone on a full-time or even a part-time basis, so you have to take what is given to you.

I have seen some organizations, mainly churches, turn to the member with a nice sound system in his/her car. And in other cases, it is the member who moonlights as a DJ on the side who gets the honor of Sound Tech. Sometimes these haphazard selections work, but most times they do not.

No matter how a candidate is selected, the organization should take an active role in ensuring that the person receives the **fundamental training** to be a Sound Tech. But they should always BOLO (be on the lookout) for other willing servants who are destined to excel in the sound ministry.

Special note to churches: Good music, sound, and visual presentation are fundamental to church growth. People are looking for a worship experience that engages all their senses, and they will continue to search until they find one. Committing resources to excel in this area helps to engage congregants so the Message may be enjoyed and acted upon.

No matter if you are a small business, church, sound tech manager, selection committee, Pastor, or staff member; the following list will give you a good start in selecting the best candidate for you:

- College Degree - a degree in music is very helpful, as the candidate has formal training, hands on experience, and knowledge of various systems. They probably also learn pretty quickly.
- Years of Experience - candidates with at least a year of experience are normally very comfortable with systems and equipment, and you can build upon that experience with formal training.
- Formal Training Outside College (On-the-Job-Training - OJT) - candidates who have significant training working under an experienced, knowledgeable sound tech are golden. They know how to use the equipment, and normally know how to interact well with others in the church.
- Good Moral Character - candidates with a good attitude, positive spirit, and strong work ethic may be trained to be excellent sound techs with time.

In most churches and small organizations, the sound tech is a volunteer. **Volunteers need to be supported** and given a chance to grow their skills. It is rewarding to see a person who volunteers for this position try hard, take directions well, and succeed. There is always room for someone whose fundamental gift and desire to be excellent is accompanied by a willingness to learn. It is up to leadership to provide the necessary resources for the training and the patience to endure the learning process.   As the church or organization grows, it is very important that the sound tech's proficiency and consistency grows with it. Some techs may disagree with me on this point, but I welcome the debate.

Almost 70% of what is takes to being a great tech is the ability to **pay attention.** I cannot overemphasize this point. Most incidents can be avoided if techs stay alert and aware as to what is going on around them, as well as what is taking place on stage. One of my pet peeves is when I'm attending an event and there are issues with the sound. I generally look back at the sound booth or under the tent and see the problem immediately: the tech texting on his or her phone, engaged in conversation with some passerby, or doing something else that has nothing to do with his or her duties. This may sound harsh, but at the end of the day you're there to do a job. If you want to carry on conversations or text on your phone during the program or event, then this is not the job for you. Individuals have paid a lot of money to put on this event, enjoy the event, and/or be a part of the event or service. Whether you have been compensated or not, it is not right and it is not fair for you to give less than your best. Your best requires you to give your full and undivided attention to the service or program.

CHAPTER 5
UNDERSTANDING THE SYSTEM:
IS IT THE SYSTEM OR IS IT ME?

Identifying and understanding your system is key. And if you have the freedom to purchase equipment, deciding on the best system for your organization's needs is critical - both from a use and budget perspective.

Good questions to ask:

> Should you have a simple system like a powered mixer board?
> Or a non-powered mixer board with multiple signal processing components?
> Possibly adding a 31 Band Eq. Power Amp with Electronic Crossover, multi-band compressors, gates, limiters and a personal in-ear monitor system?
> Maybe a Loudspeaker Management Processor?
> And lastly add a Digital Mixer which mostly has all of the above components already built-in?

You also need to know exactly how much power the sound system needs to function correctly. IT MATTERS. There's nothing worse than purchasing thousands of dollars of equipment only to find that the building or the venue cannot support it electrically. It is then you realize the upgrades that you have just purchased were for nothing.

In some cases, sound techs are put in a situation that they neither have knowledge or understanding of the system that they are to control, and they do not realize it until the day of the event or service. This is a serious no-no! It happens when a person is brought on board after the system was installed and/or joins the team fresh. At the end of the day, it does not matter why; the bottom line is that this is your baby now and you are going to have to burp it: you have to educate yourself on the system and operate it like you were born sitting at the wheel.

As the Sound Tech, you must know the basic signal flow of the system **and** have the willingness to learn the capabilities of the total system. The total system is everything that is needed for the event or service to look, sound, and be excellent. This includes signal, flow, microphone assignments, the mixer board, and amps.

Basic Live Sound Setup

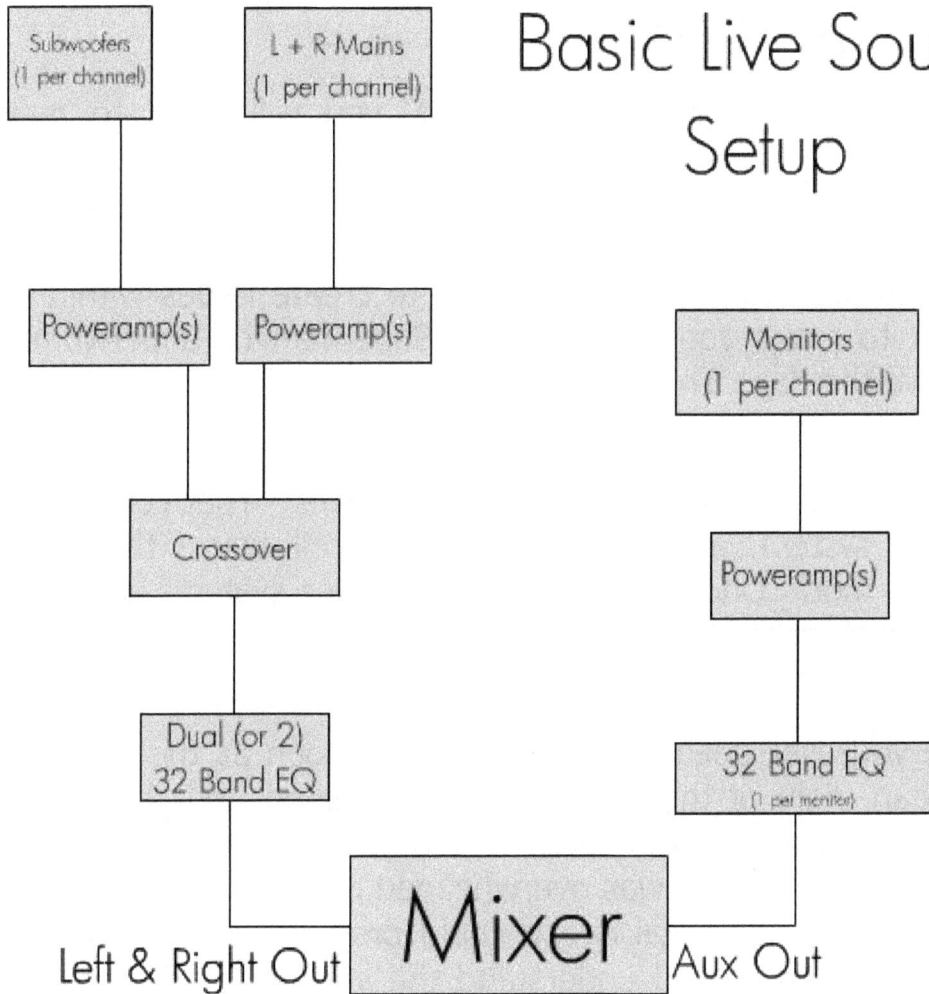

Subwoofers (1 per channel)	L + R Mains (1 per channel)

Poweramp(s)	Poweramp(s)

Crossover

Monitors (1 per channel)

Poweramp(s)

Dual (or 2) 32 Band EQ

32 Band EQ (1 per monitor)

Mixer

Left & Right Out   Aux Out

After about three months, an experienced Sound Tech should know the system flow, or have redesigned it to best address the organization or church's needs. The kinks should have been worked out, and any troubleshooting resolved. (And for the record, troubleshooting is not shutting the system down and/or rebooting.) And at 90 days in, you should know the system well enough to know when something is not right.

As you become more experienced, you will learn to make adjustments to the system so quickly people will not notice there was an issue in the first place. Speed in this instance is critical in keeping the focus on the program or service, and NOT on how the system is (or is not) performing. You do not want to create issues within the system just to make yourself look like Superman or Superwoman, ready to save the day. This is not about you.

Most skilled techs and operators run into problems often. But they are fixed and handled before anyone notices. That comes from knowing the system, and it does not matter if you - the tech - installed it or not. Once you get to know your system, and get it functioning well, people will come up to you and have nice things to say (instead of complaints).

Have you ever been to a venue and as soon as the band and singers began, right at the height of the first song, the power went out? It happens more often than it should, and many times it is because either the system was overwhelmed or required more power than the building's electrical circuit could produce. Why? Because the tech did not know and understand the system (or did not take the time to check the electrical room and access the power capabilities). Putting too much pressure on the electrical system will cause a shutdown, trip a breaker, or blow a fuse. This is especially common in older venues that have fuse boxes or outdated electrical service - half the building loses power.

Sound In Action - Real World Example

About 15 years ago, I was hired to do sound for an event on a college campus. We got in that morning to do setup. Because this venue was a fairly new building, it never occurred to me to check the electrical room for any issues (something I normally did for older buildings). There were a lot of people, vendors and assistants, moving around during sound check and setup. We were so busy, I never got a chance to drive the system real hard. Big mistake, and one I never made again.

During the first song, the power went out in the entire building. The building supervisor came rushing to the back to reset the electrical circuit. The power came back on. The supervisor came over to me and said, "Yeah, we've been having this problem. We don't know what's really going on." I have a little bit of electrical knowledge, and knew exactly what the problem was. I told the supervisor that the lighting, power, par cams, power screens, and projectors were running too much on one phase. Thankfully, I also knew what to do to fix it: we quickly pulled all three of my power amps out of the rack and plugged them in separately, one on each wall. As a result, each amp pulled from a different circuit breaker and no longer overworked the single 15 or 20 amp circuit breaker.

For the techs that aspire to have their own sound production company, it is vital to get some form of low voltage knowledge training. You will definitely need it. The electrical output may vary with any (and maybe every) venue that you will be asked or hired to operate sound in. I have seen techs overcomplicate and overtax their systems, causing severe damage (and sometimes destruction beyond repair). How? By using outdated or incompatible equipment.

A word of caution: please purchase equipment that is conducive to the vision of the programs you currently have. As the vision (and resources for it) grow, you can add other equipment and capabilities later. A system should reflect the vision (or ministry in religious settings). For example, if you are in a ministry and you are the only sound tech or volunteer, then it does not make sense to have a system that requires three or more people to man it. The ministry cannot be held hostage because you cannot make it or get a "call out" from someone else on the team. The system should be set up to where the key people in the group (business owner, Pastor, Minister of Music, or select designee) can take it and run. Remember - the "show" must go on.

Also, when it comes to purchasing equipment, keep in mind to KISS: keep it simple students. I have been called out to do service calls and the system had so much equipment integrated into it that it was hard to see where to start. There were so many different processors, many of them were not hooked up correctly, and others were not even needed.

CHAPTER 6
VALUE IN RELATIONSHIPS

In any setting, but particularly religious organizations, all teams must realize that they are working in their gift or calling - just as is their leader (or pastor). Therefore, when it comes time for a judgment call or a key decision, everyone must defer to the leader. That is simply how it works.

When you are in a setting or room full of strong thinkers, there are often disagreements. But remember: the final decision rests with the leader. As such, you must put forth 100% effort in the quest of the vision. Trust me, a good idea will reveal itself over time and so will a bad one.

Sound in Action

When I first began doing sound, I was put in charge of the audio and media ministry at my church. The media ministry had fallen so far off the radar that the tape/CD duplication sales were almost non-existent. We had a big event coming up and I made a proposal, a plea really, to the pastor to upgrade the church's equipment and purchase a CD recorder and duplicator for this event. Because the media ministry bar had been set so low, and so many tapes had been returned due to poor quality, my pastor's response was exactly what I figured it would be: no. Specifically, he said, "I'm not doing it. I'm NOT putting any more money into that Ministry."

For two weeks I badgered, pushed, prodded, and basically became a nuisance to change my pastor's mind. He would not budge. So, I took it upon myself to go out and rent a CD recorder, and I recorded the first night of the event. That first night went so well, I was filled with joy. There were so many requests for CDs that we were almost overwhelmed. (Almost!) We sold over $200 in CDs by the end of the night! Since we did not have a duplicator, it took me four and a half hours to create 30 CDs.

When Pastor came in the next day and found out how much the church had made with the CD sales, he was stunned - but only for a moment. He stated, "let me hear what it sounds like." His concern was the quality of sound. Although $200 is a relatively small number, when you are talking about tapes and CD's, it says and means a lot.

A man of great wisdom, after listening to the CD, Pastor immediately asked me, "How fast can you get a duplicator here, and how much did it cost to rent the recorder?" Sometimes it takes showing better than telling.

Sound in Action

Once when I was over the sound department at a local ministry, the pastor came to me with a vision: he wanted all cordless microphones. I did not agree with that vision at all. At the time, there were not that many people on the praise team and there was nothing really wrong with the corded mics that were in use. I felt that spending money on all cordless microphones would be a waste of resources. The pastor was adamant that the cordless mics were necessary. He said, "The praise team is coming! It's coming, don't worry about it, just get me a price on all cordless microphones." I did not agree with the vision, but I did as I was asked. I am glad I did because he was right: the praise team did grow and the ministry is doing well.

One of the greatest contributions a sound tech or operator can make to the overall vision is the ability to rationally put things into proper perspective, keep the peace, and solve problems they arise.

Realizing this charge and developing your ability to do it well is fundamental to your success. Make every effort to get along with every level of the organization: the leader (pastor), management (director of music), staff, event planners (worship team), and congregants (audience). When you get along with people, it makes it easy for them to defer to your knowledge and experience - no matter how limited or great it may be. This is important because many people - particularly those in religious settings such as pastors and ministers of music, have little or no knowledge of the technical aspects of sound, lighting, or video.

Always keep the lines of communication open. If there are things you want to communicate to the leader(s) of the organization or church, be respectful and work to pick a time that is convenient for them. Trying to talk to business immediately after the event or Sunday service sometimes is not wise: EVERYONE is trying to do the same thing. Instead, set a day during the week to have dialogue with them about the system, **address issues, and resolve concerns**. And if it is a tough conversation, think about doing it over a meal like lunch.

You are probably thinking "what does this have to do with sound?!" Being able to do a good job is all about relationships. Being respectful of others, their time, and their needs helps to cultivate and strengthen good relationships. When you have a good relationship with someone, you are able to have tough conversations, and be open to constructive criticism - giving it and receiving it. One of the greatest challenges that I have had as a sound tech is being open to constructive criticism. I am better about it now, but it is a work in process.

You must also be flexible. Don't ever be afraid to admit that if you don't know, you just don't know. As the sound tech, you should do your best to accommodate the leader's request. Now, of course it must be appropriate, within reason, and within your ability to do so. Even if you strongly disagree with the request, you should give it appropriate consideration. None of us is right all the time.

Many times in church, and even in the secular world, problems are a result of poor communication skills. If you know that your communication skills need work, then have someone you know and trust help prepare you for difficult conversations. At the same time, take some training classes to improve your skills. Toastmasters is a great speaker training organization.

If at all possible, never leave unresolved issues "lying around". It is a breeding ground for malcontent and could possibly set up an atmosphere of negativity and poor productivity. We have to remember that this is not the "Me, Myself and I Show". We all must work as a team. To put it simply, we must COMMUNICATE! I like what Dr. Gary Taylor Sr. Pastor of Open Word Christian Ministries, Inc., has to say on this topic:

> *Revelations 2:7 states 'he that hath an ear, let him hear what the spirit saith unto the churches'.*
>
> It is very important for sound techs to understand how they can make or even break a service. In my dealings with sound techs over the years, I have found that most of them don't understand how they can impact the service in a positive or negative way. Sound techs have the ability to help the service or make a mess out of it, from feedback, to the volume being too loud, etc.
>
> From a pastor's perspective, in my dealings

with sound techs is that I find so many of them don't really understand the role that they play. When the sound is too loud, for example, it causes the service to be unbearable. People in the church find themselves not listening, and even leaving because the volume is too loud.

Or the sound tech didn't record the service, or if they did, the volume was too loud which causes distortion on the recording. [That] can cause a person to return it. Also, the person can't hear the Word that they need for their life.

Just like the minister should be skilled, the sound tech should be skilled as well.

Both the pastor/minister [and sound tech] should have a good working relationship which will cause the service to run smoothly without any hindrances.

CHAPTER 7
TOO MUCH ON THE LINE TO FAIL

As a sound tech or sound operator, you and the leader of the organization are never going to agree 100% all of the time. But you should always strive to keep a harmonious balance in the leadership at all times. Although challenging, your role is an important one and can be very rewarding if you have the right attitude.

Having the right attitude allows you to grow, and not just survive but to thrive. Be open-minded, develop a "thick skin", and enter every venue with joy in your heart - this applies whether it is a church or a secular organization. No one wants to work with a loud, angry, irresponsible sound operator or sound tech.

A lot of really bad communication techniques are born out of good intentions. Your appraisal of a situation could be right, but if a situation is handled poorly, your righteousness could result in damaged hearts and lives. Once again to quote Whitney George, "You won't be judged by those situations you've been given, but rather by the way you handled it." Here is what Bishop Terrell Beard

Pastor of Enduring Faith Ministries, Inc., has to say about the impact the sound tech's attitude has on the worship experience:

> An important and necessary component of contemporary worship services is a skilled, spirit-filled sound operator. This individual must have a harmonious relationship with the senior pastoral leadership of the church in order for things to go as visualized through the Spirit of God.
>
> In the Black Church, many times ministerial needs are not met nor deliverance achieved due to non-existent or malformed relationships between pastors and music, sound, and/or worship leaders. Hopefully, that gap can be bridged in part through this guide and accompanying workshop, and a commitment to future excellence.
>
> In the past, the lack of specific knowledge and the availability of adequate equipment put minority congregations at a distinct disadvantage when it came to presenting a marketplace quality product into the public domain. In many cases, upgrading was seen as back sliding. As a result, the growth in predominantly small churches was stunted and delayed as they struggled to understand why members were leaving for more progressive ministries.
>
> There were a number of reasons for the exodus we witnessed, with structure and

organization topping the list. This fact is relevant because those leading this move appear to be the next generation of leaders – the youth. The next generation has a major role to play in the shaping of worshipping arts ministries. The shifting paradigm in the music ministry is requiring a philosophical approach that can only take place with the infusion of new creative input from the next generation. If we allow for an environment conducive to sincere worship to be created, then, and only then, can true deliverance be obtained.

We must come to understand that the next generation is our legacy and as such represents our investment to Kingdom building, which we cannot afford to release to those who made no resource contribution. It is my hope and desire that we as senior leaders can seek God's guidance for direction as we move to minister to an ever increasingly complex body of believers.

CHAPTER 8
CONCLUSION

You are human. You will make mistakes. Learn from them. Commit to continuing education and training. Ask for help.

Failure only comes when you allow your attitude to create or contribute to a negative environment. Stay away from negative conversations that simply "stir the pot" rather than move things forward and upward. Keep your eyes focused on the "big picture" - a meaningful experience for the audience or congregation - do not sweat the small things. Share and talk in a way that shows your genuine commitment to excellence.

Be an example of Christian love; your actions promote Unity and Harmony. Remember, there is no I in team.

ABOUT THE AUTHOR

Henry Tucker was born the first of December to the late Norman H. White and the late Alfreda Tucker. He has four sisters, three brothers, and a host of shared Godparents. At the age of seven, Henry was baptized at Bethlehem Missionary Baptist Church under the direction of Rev. Charles Perry. He was born with a love of music in his heart. Out of nowhere, Henry learned to play drums. His mom saw that his love for gospel music kept growing, and she put him in the company of anointed musicians where he began to develop his gift. This same gift has opened so many doors for Henry in his life. He has had the opportunity to play for some of the most talented and professional musicians in the country.

Henry is a graduate of Lake Weir High School in Chandler, Florida. After high school, he began his life career path. Henry has always had a love for truck driving, most likely because his dad and all of his uncles were truckers - guess it must be in the blood. After two years delivering home appliances, Henry transitioned to serving as a correctional officer for three years. After this time, he went back to driving trucks.

Because of Henry's love for music - not just gospel, but all music - and a passion for making good music sound better, he was elevated to sound tech and head of the media ministry in 2002 at Greater New Bethel under the leadership of the Late Dr. Dennis Hagins, Sr. Under Dr. Hagins' guidance, Henry was able to work with Skip Winseck, a sound engineer with 30 years of experience. Skip trained him and taught Henry all of the fundamentals of sound without charging him or the church a dime. The only cost Henry incurred was time to continue to learn, and then learn some more.

In 2004, Henry started his own small business called Open Sound Productions. He began with service-calls to churches, but then moved to training the churches' personnel. Then, the company began providing sound for concerts, revivals, and plays. Because Henry had a love for helping small churches and helping people who have a heart to learn (and he grew weary watching so many churches get taken advantage of buying equipment that did not suit their ministries), Open Sound Productions started consulting services to ensure every church and business that had utilized its services were educated on all of their equipment.

In 2011, Henry moved to Atlanta and opened Tucker's BBQ, LLC, a mobile catering company. Henry credits Bishop Terrell Beard, the owner of Terrell's BBQ and Pastor of Enduring Faith Ministries in Gainesville, Florida for inspiring him to start this new venture. (Henry had the honor of being a member and serving as the head of the Sound and Media ministry for several years at Enduring Faith.) Since opening, Tucker's BBQ has catered and served family reunions, graduations, and networking parties from Ft. Lauderdale, Florida to Sacramento, California, and from Detroit to Philadelphia. And in January of 2016, Henry started HT Logistics, LLC trucking company.

Currently, Henry serves as head of the audio ministry at Open Word Christian Ministries in Fairburn, Georgia under the pastor Dr. Gary K. Taylor Sr.

"I have gained so much wisdom and experience in organizing and coordinating workshops, concerts, trainings, and running multiple businesses. I believe that with all that has been given to me, I cannot be selfish and keep it to myself. I have a different calling than standing behind the pulpit; my ministry is to help spread the word in a different form. This is my lane, and I drive it well."